ROAD ROLLERS

by Katie Chanez

Cody Koala

An Imprint of Pop!
popbooksonline.com

abdobooks.com
Published by Pop!, a division of ABDO, PO Box 398166, Minneapolis, Minnesota 55439. Copyright © 2020 by POP, LLC. International copyrights reserved in all countries. No part of this book may be reproduced in any form without written permission from the publisher. Pop!™ is a trademark and logo of POP, LLC.

Printed in China
052019
092019

THIS BOOK CONTAINS RECYCLED MATERIALS

Cover Photo: iStockphoto
Interior Photos: iStockphoto, 1, 5, 7 (bottom left), 8, 12, 14, 15, 18; Shutterstock Images, 7 (top), 7 (bottom right), 9, 11, 17, 21

Editor: Meg Gaertner
Series Designer: Sophie Geister-Jones

Library of Congress Control Number: 2018964599
Publisher's Cataloging-in-Publication Data
Names: Chanez, Katie, author.
Title: Road rollers / by Katie Chanez.
Description: Minneapolis, Minnesota : Pop!, 2020 | Series: Construction vehicles | Includes online resources and index.
Identifiers: ISBN 9781532163357 (lib. bdg.) | ISBN 9781644940082 (pbk.) | ISBN 9781532164798 (ebook)
Subjects: LCSH: Road rollers--Juvenile literature. | Steam rollers--Juvenile literature. | Construction equipment--Juvenile literature. | Construction industry--Equipment and supplies--Juvenile literature.
Classification: DDC 625.8--dc23

Hello! My name is

Cody Koala

Pop open this book and you'll find QR codes like this one, loaded with information, so you can learn even more!

Scan this code* and others like it while you read, or visit the website below to make this book pop.

popbooksonline.com/road-rollers

*Scanning QR codes requires a web-enabled smart device with a QR code reader app and a camera.

Table of Contents

The Road Roller Can Help!

Construction workers are building a new road. They lay down **asphalt**. The asphalt is bumpy. A road roller drives over the bumps. It makes the new road flat.

Watch a video here!

A Road Roller's Job

Road rollers are used to **compact** surfaces. They make surfaces flat. Usually, road rollers help build roads. But some road rollers help flatten trash in **landfills**.

Learn more here!

Roads are often made of **asphalt**. Construction trucks pour asphalt onto the ground.

Road rollers drive over the asphalt. They make the asphalt flat. The new road becomes smooth.

Chapter 3

Parts of a Road Roller

Road rollers have several parts. The big front wheel is called the **drum**. The drum is very heavy. When it rolls, it presses down the ground beneath it.

Complete an
activity here!

The cab is behind the drum. The cab is where the driver sits. The driver can control how quickly the drum spins. The driver can also **steer** the road roller.

Some road rollers have wheels in the back. Others have a second drum instead.

The second drum goes
behind the cab.

Some rollers have sprinklers that wash
the drum after a job is done.

Chapter 4

Types of Road Rollers

Some road rollers have **drums** that **vibrate**. The vibrations cause the bits of **asphalt** to move even closer together. The road roller can **compact** more quickly.

Learn more here!

Some road rollers don't have a drum at all. Instead, they have four or six tires. These rollers are usually used for light soils, such as sand. They also smooth the tops of finished roads.

Some road rollers have drums that are not smooth. The drums have "feet" that stick out several inches. The feet help compact thick soil.

People have been using road rollers for more than 100 years. The first road rollers were powered by steam.

Making Connections

Text-to-Self

Have you ever seen a road roller? Where was it? What was it doing?

Text-to-Text

Have you read books about other construction vehicles? How are they similar to a road roller? How are they different?

Text-to-World

Flattening the ground is often the first step in many construction projects. What other steps are there?

Glossary

asphalt – a sticky black substance used to build the surfaces of roads.

compact – to flatten or squish something.

drum – the round part of a road roller that rolls across surfaces to flatten them.

landfill – a place where trash and other waste is brought and buried.

steer – to guide or control movement.

vibrate – to move back and forth very quickly.

Index

Online Resources

popbooksonline.com

Thanks for reading this Cody Koala book!

Scan this code* and others like it in this book, or visit the website below to make this book pop!

popbooksonline.com/road-rollers

*Scanning QR codes requires a web-enabled smart device with a QR code reader app and a camera.